Horatio Robinson Storer, Albert Day

Methomania

A Treatise on Alcoholic Poisoning

Horatio Robinson Storer, Albert Day

Methomania
A Treatise on Alcoholic Poisoning

ISBN/EAN: 9783744670302

Printed in Europe, USA, Canada, Australia, Japan

Cover: Foto ©berggeist007 / pixelio.de

More available books at **www.hansebooks.com**

METHOMANIA:

A TREATISE ON

ALCOHOLIC POISONING.

BY

ALBERT DAY, M.D.,

SUPERINTENDENT AND PHYSICIAN OF THE WASHINGTONIAN HOME, BOSTON;
AND FELLOW OF THE MASSACHUSETTS MEDICAL SOCIETY.

With an Appendix,

BY

HORATIO R. STORER, M.D.,

PROFESSOR OF OBSTETRICS AND MEDICAL JURISPRUDENCE IN BERKSHIRE
MEDICAL COLLEGE.

BOSTON:

JAMES CAMPBELL,

MUSEUM BUILDING, 18, TREMONT STREET.

1867.

CAMBRIDGE:

STEREOTYPED AND PRINTED BY JOHN WILSON AND SON.

PREFACE.

In offering this little volume to the public, I do not intend it to take the place of more elaborate works on Alcoholismus, which have from time to time been given to the world by the best writers on medical science. These few pages are suggested by the experience of more than ten years' constant observation of the effects of alcohol upon the human economy. I have had under my care over two thousand cases of inebriety, usually in its worst forms ; of this number, more than four hundred suffered from what is called *Delirium Tremens*, all of which cases have been under my own observation from the beginning of the attack to the end. I have, for the love of my race, observed with critical care the effects of intoxicants upon the mind ; and I cannot refrain from raising a warning voice, feeble as it may be, against a habit which permeates all society, from the highest

[3]

to the lowest, and which will continue to entail upon society its legitimate fruits of drunkenness and crime.

I have intended this work as a suggestion to the public generally, as well as to the medical profession. I hope at least it may fall into the hands of the young, that they may guard themselves with care from any habit that can possibly lead to a disease so fatal to mind and body.

I do not claim to have exhausted the subject in this little volume: it is far from perfect; it assumes to be but a handful gathered from a vast field of observation. It is compiled amidst much care; and this must be my excuse, if imperfections are detected by the critical eye.

Washingtonian Home, 887, Washington Street,
Boston, Jan. 1, 1867.

METHOMANIA.

I HAVE selected this title as an appropriate general name for that disease which, in its several forms or stages of development, is variously termed Drunkenness, Inebriety, Dipsomania, Methexia, &c.

Causes. — The simple form of this disease results from the introduction into the system of any of those poisons which first stimulate, then narcotize.

The chronic form has often causes more remote than continuous stimulation, under the action of which the passion for stimulants amounts to a *mania*, seemingly irresistible and incurable. These causes are either idiopathic, or are induced by habits or manner of life. Among these may be mentioned a

[5]

mental constitution unbalanced in its moral faculties; congenital physical weakness, resulting in a morbid tendency to melancholy; a weak individuality; and a disposition easy, good-natured, and social. The disease is frequently induced by misfortune, which darkens and embitters life; or by contrary successes, which unduly excite the mind. The miserable, the irresolute, the listless, the unoccupied, and those too much burdened with care or labor, are the subjects of it.

As the physician is more frequently called to deal with the morbid effects produced by alcoholic poisoning than by any other mode, I shall confine myself to its discussion.

The symptoms of alcoholic poisoning in its early stages are, usually, great mental exhilaration or excitement, with frequent and strong pulse, animated features, increase of appetite and thirst, and of the secretions of skin and kidneys. Under the development of individual characteristics, also, it may be seen that the sad become sadder; the light-hearted, gay and boisterous; and the ill-tempered, quarrelsome and vindictive. As the poison is increased, ideas become confused, and the reasoning powers so disordered as to produce a condition approaching delirium; the sensorial apparatus is disturbed, causing a thickness of utterance, double vision, and the loss of voluntary control of muscular movements; vomiting sometimes ensues, but more frequently a deep sleep, which continues as long as the influence of the poison lasts. This condition varies from that of a profound sleep, with no unusual phenomena of disturbance in the pulse, breathing, appearance, &c., and from which the patient may be aroused to sensibility, — to a comatose state, or to a torpor resembling apoplexy, the face unnatural, either pale or flushed; the eyes vacant, and sometimes glazed; the pupils much dilated, and little affected by light; the head hot; the temperature of the extremities low; the pulse feeble and decreasing, sometimes entirely wanting at the wrist; and the respiratory movements infrequent, and often laborious and convulsive.

These symptoms are sometimes accompanied, in the more advanced stages, by Strabismus, general tetanic convulsions, or spasms of particular parts, which are unfavorable signs, and frequently indicate a fatal termination. Appearances after death very nearly resemble those produced by asphyxia. The face is livid or swollen, bearing marks of convulsions; the eyes are prominent, and the pupils dilated; dark fluid blood is found in the cellular tissue and lungs, the latter being much expanded by frothy mucus in the air-cells. The right ventricle of the heart, with the pulmonary arteries and systemic veins, are found filled with blood; while there is but little in the left ventricle and arterial system, and that of a dark color. The liver, spleen, and kidneys are loaded with venous blood; and blood of the same character fills the sinuses, veins, and even the smaller vessels of the encephalon. The substance of the brain is usually white and firm, as though it had been steeped in alcohol for an hour or two. Serum is found in the ventricles and between the membranes;

and, in some cases, the fluid in the ventricles is found to have all the qualities of alcohol, or is at least strongly impregnated with the same. The stomach varies but little from its normal state, except in cases where the habit of using alcoholic stimulants has long existed. The mucous coat is then usually found to be thicker, softer, and more vascular. The smaller intestines are similarly affected, sometimes through their entire length, and are often injected and ulcerated. In some rare cases, this mucous coat is found hardened ; in fact, the alimentary canal takes on various forms of disease, according to the temperament, habitation, age, and habits of the patient. The bodies of confirmed drunkards are rapidly destroyed by putrefaction, and can be easily distinguished from the bodies of those who die of other diseases.

While severe sickness, and even a fatal termination, will sometimes ensue as a direct result of other forms of disease, of which, to be sure, this is the primary or exciting cause, the great evil of this poisoning is, that its

1*

effects may be called agents for the production and development of chronic organic diseases.

The first, which owes its existence directly to alcoholic poisoning, and which seldom, if ever, in its distinctive symptoms has any other cause, I will call by the general name of *Mania à potu.* It has been customary to apply the term *Delirium Tremens* to all manner of delirium resulting from the excessive use of alcohol. The public generally, and even many medical practitioners, have failed to make a proper distinction between that form of mania which may be justly described as *Delirium* with *tremor*, and those other forms which may be included under the general title of *Mania à potu.*

It is generally conceded that cases properly diagnosed as delirium tremens are rare, and usually occur in persons who lead a sedentary life, and are of nervous or sanguine temperament; and in those whose intemperate habits have been long confirmed, and uninterrupted by occasionally violent and unusual excesses.

The distinctive symptoms of this stage of
mania are, in its first developments, anorexia,
insomnia, great tremor of the muscles, especially the tongue, with constant anxiety, and
apprehension of danger. A further development is marked by peculiar mental aberration,
characteristically diversified in every case.
The patient becomes the victim of distressing
and frightful illusions, the most common of
which is the presence of imaginary animals,
as dogs, cats, rats, snakes, &c.; noises are
also heard, and absent persons addressed as
if present; or he is surrounded by enemies
or pursued by them, is required at some other
place on important business, &c. He is con-
tinuously active, vigilant (hence the disease
is by some called Delirium Vigilans), in
constant motion, and requires to be closely
watched. This state of mind is accompanied
by tremor of the muscles; a more or less disordered state of the digestive apparatus;
anorexia continuing; the stomach frequently
refusing to retain any kind of nourishment;
the tongue furred; the bowels usually consti-

pated; skin cool, often moist; and the pulse feeble and thready.

Insomnia continues for a time; but return to sanity generally follows the first or second sleep. The prognosis is usually favorable. Death rarely ensues from this cause alone, unless several severe attacks have been previously experienced. A fatal termination often results from the co-operating influences of other disease.

In cases occurring after wounds and surgical operations, or where other morbid affections, as pneumonia, &c., are evident, the prognosis is more unfavorable.

Where insomnia continues for several days, accompanied by great nervous exhaustion, continued delirium, feeble and rapid pulse, increasing tremor, fitful excitement, and struggles followed by profuse and cold perspiration, the general chilliness of the surface increasing, the prognosis is serious, and a fatal termination probable.

Death at this stage sometimes ensues from convulsions, but more generally from exhaustion immediately preceded by a calm, and the patient dies comatose.

Another form of mania à potu may be termed *Delirium Ebriosum.* This more frequently results from a paroxysm of intemperate drinking than from habitual intoxication, and resembles in character that of the preceding stages. The patient, however, is less given to hallucinations, less troubled with frightful phantasms ; and although in some instances violent, and even ferocious, is more generally good-natured and docile, sometimes gay and hilarious. This form is usually unaccompanied by tremor ; or, whatever of tremor may exist at first, speedily disappears on the development of active delirium. The head is generally hot, the face flushed, the pulse full and frequent, with throbbing carotids, febrile movements, and other symptoms indicating cerebral congestion. Insomnia is not often long continued ; and, as sanity follows sleep, the patient rallies rapidly. This is the most common form of mania by alcoholic poisoning.

Other degrees exist, in some of which the

patient is in a state of partial consciousness, or may have lucid intervals; in others, he seems entirely conscious, but is subject to illusions of sight or hearing, which he knows to be delusions, but cannot dispel from his mind. Such wanderings and illusions rapidly disappear with the elimination of the poison.

Another form of mania may be called *Hysterical Tetanus*, in which the patient labors under constant anxiety and apprehensions, with occasional attacks of tetanic spasms. I have taken the liberty of improvising a name for this form, having observed a large number of cases, in which the above symptoms were very prominent; and the term Hysterical Tetanus better describes the disease than any other known to me.

This disease is liable to continue two or three days, and is attended with more or less disturbed sleep; but rapidly disappears on a discontinuance of the poison.

Insanity. — All competent authorities are united in attributing to alcoholic poisoning a large proportion of the numerous cases of settled mental derangement, although opinions vary somewhat as to the relative proportion between this and other causes. It seems probable that insanity is often attributed to the immediate exciting cause, while the remoter and producing cause or causes are either unknown or overlooked.

As already stated, a large proportion of the disorders known to medical practice are capable of being either induced or developed by a free and constant use of alcohol; and, as a state of insanity frequently succeeds or accompanies many of these disorders, a thorough diagnosis, involving a history of the patient, a close scrutiny into preceding circumstances, would more frequently than is done identify alcohol as the predisposing cause, — the great disturber. This poison may be said to act directly on the nervous centres, or at least its presence in the blood has a peculiar tendency to disorder the functions of the nervous system; and it is noticeable that the encephalic portion seems to be the special field for its disturbing chemical action. An affinity seems to exist between alcohol and nervous matter, which may account for its special power in deranging the nervous system. As strychnine acts directly on the spinal cord, so alcohol has perhaps its most direct effect upon the encephalon. It has been said by competent authority, that in all diseases of the encephalon, except those which are purely constitutional, the chances are largely in favor of the supposition that the patient has been intemperate. There can be no doubt, that the constant use of alcohol gradually modifies and disturbs the nutritive processes of the brain, and that excessive indulgence actually alters the vital properties of the organ ; so that the various exciting causes which result in insanity, while they would fail of effect on the healthy brain, find the ground well prepared for their influence, and are almost sure of success in a brain whose nutrition is defective, and therefore weak and disordered. Experience embodied in statistics establishes the fact, that alcohol is responsible for a large share of the cases of mental derangement; and I believe, that, taken in all its relations to diseases of the mind, it is a more prolific cause than is generally supposed.

Inflammatory Diseases of the Brain. — When we discern the seeming affinities of alcohol by its effects on the brain, we are prepared to acknowledge that it may be the immediate agent in producing such dangerous, though at times temporary, inflammations as Cerebritis, Meningitis, &c. ; or at least that its tendency is to prepare the brain for their production by other and more immediate causes. Medical writers universally agree that such results are to be anticipated. These diseases are often mistaken in diagnosis for the mania above mentioned, where the previous habits of the patient are known to have been intemperate. The distinctive symptoms, however, of meningitis, and by which diagnosis would distinguish it from mania à potu, are usually a tense and hard pulse, dry and feverish skin, severe headache, absence of tremor, a peculiar mania different from that of mania à potu, constant vomiting, labored respiration, and great nervous prostration. These symptoms rarely occur in mania à potu, except in extreme cases and in the last stages; but it seems probable, that in many, if not all, of the fatal cases of the delirium of drunkards, in its several forms, the disease culminates in severe inflammation of the meninges. It has been asserted, that the excitement produced by a single alcoholic poisoning would pass into meningitis, but for the extraordinary efforts of the system to eliminate the poison by exhalation from the lungs, and increased secretions of the skin, kidneys, &c. In cases where this elimination is rendered abortive by a constant re-introduction of the poison, it is not at all impossible that meningitis may supervene as a direct result: indeed, I have long since arrived at the conclusion that this occurs very frequently.

That apoplexy, or what is known as cerebral apoplexy, would probably be induced by alcoholic poisoning, may be presumed from what has been said of the relations of alcohol to the brain. A state of deep intoxication closely resembles apoplexy, but may be distinguished from it by the stertorous breathing, and slow, irregular pulse of the latter. The limbs, too, are frequently paralyzed in apoplexy: this is not witnessed in mere intoxication. The strong smell of liquor, also, which taints the atmosphere around the intoxicated, is a marked diagnostic point. While it may be doubted that apoplexy ever directly results from alcoholic poisoning, there can be no doubt of the latter's agency as a predisposing cause. Da Costa observes, that the hemorrhage could be detected with greater certainty, were it not that the extravasation so often takes place in an *already diseased brain.*

It is well known that the condition of *Plethora*, which usually follows excessive eating and drinking, is quite favorable to apoplectic hemorrhage. Alcohol, by impeding the circulation, thus preventing the assimilation of solid food, and, as it were, retaining it by force in the stomach, tends, more than is supposed, to the same result. It might be gradually induced by the use of this stimulant, although never carried to the extent of absolute intoxication, and then would most likely be attributed to gluttonous excess in eating, and the remote cause — alcohol — be overlooked. Paralysis, or the cerebral form of paralysis, depends on nearly the same conditions which result in apoplexy (of which it is the frequent accompaniment, precursor, or successor), and, for the same reasons, may be said to be superinduced or accelerated by free use of alcohol. In cases of paralysis of particular parts, as in palsy or paraplegia, the most conclusive reason, perhaps, for considering alcohol as an active agent in its production is, that a total abstinence from alcoholic drinks will frequently be followed by a complete restoration, or almost always by a very sensible relief.

Epilepsy, which is generally attributed to disordered nutrition, is very likely to arise from the same cause; and mitigation of its distressing symptoms generally follows the discontinuance of the inciting cause. The results of this disease — softening, dementia, decline of mental vigor, weakened memory, and loss of nervo-muscular energy — may frequently be traced to alcoholic poisoning.

The most marked effect of alcohol, as a direct or indirect agent in the production of those forms of encephalic diseases which are attributed to its use, seems to be the impairing of nutritive functions. The brain especially is thereby exposed to attacks of disease from various pathological conditions of other organs. Alternate excitations and corresponding depressions weaken ; and these, constantly repeated, produce chronic disorder.

The stomach seems to be less strikingly affected by this poisoning than the brain, and some other parts of the system ; but there is abundant reason for believing, that the excessive use of alcoholic stimulants will, even in a single debauch, produce inflammation or morbid changes in the mucous membrane. This was observed in the case of St. Martin, after a free use of ardent spirits, although no symptoms, local or general, indicated the true condition of the membrane.

The effects of a small quantity of alcohol is to quicken the circulation, and increase the functional activity of the organ ; but, when the quantity taken is large, all the effects of a strong irritant follow, — diminished functional activity, congestion or severe inflammation. Habitual irritation results in an altered state of nutrition, which, in the *post-mortem* of drunkards, is indicated by the thickened and softened membrane.

There is reason also to suppose, that an ulcerated condition of this organ sometimes results, which cannot be mitigated except by total abstinence from the use of the irritant which is the active cause.

The serious derangements of the general nutritive functions lead naturally to that form of dyspepsia known as Gastritis. This disease is found by medical men to be common among those who make frequent use of alcoholic stimulants ; and the disordered state of the mucus membrane, as before stated, is not limited to the stomach, but sometimes extends through the entire length of the alimentary

canal, as is sufficiently indicated by inflammation, ulceration, hemorrhage, &c.

While the agency of alcohol in large quantities is admitted to be very great in the production of gastric diseases, it is a too general
opinion, that the *regular, habitual* use of it in
small quantities acts favorably by the gentle
stimulation of the digestive functions.   The
grounds for entertaining this opinion are falla-
cious ; and I agree with Dr. Hodgkins (Cope-
land, p. 784) in saying that " the habitual
employment of such stimuli must be injurious,
by blunting the sensibility of the stomach to
those articles which are really nutritious, as
well as by contaminating, by the admixture
of a deleterious principle, the nutritious
juices which the absorbant vessels have to
imbibe."

Lungs. — Alcohol reaches the lungs through
the medium of the circulation ; and, as so large
a portion of it is eliminated by exhalation
therefrom, there follows, inevitably, an irritation of the delicate membrane lining the air-
cells of the lungs, and the mucous surface of

the bronchi and trachea. It is natural to suppose, that impeded respiration, coughs, and even tubercular disease, would follow constant irritation. The truth is, that not only does this stimulus predispose the lungs to disease, but it also acts, in many cases, as the direct, the immediately exciting, cause of consumption. The best observers are agreed in this.

An enlargement of the heart (Hypertrophy) is frequently observed in the *post-mortem* of inebriates, with ossification of the valves and arteries.

Liver. — Probably the liver is more liable to disease from alcoholic poisoning than any other single organ. Stimulated to increased activity, it eliminates some portion of the alcohol in the blood transmitted through it; and the contact of this poison with its tissues predisposes to various disorders. It is unduly burdened and overworked by the retention of much hydro-carbonaceous matter, which, were it not for the presence of alcohol, would be carried in the blood to the lungs, and there thrown off in the process of respiration. And, if this labor be prolonged, it results in chronic weakness, and consequent liability to disease.

An acute and chronic inflammation of the liver is indeed a frequent disease, resulting from long or continued use of spirits or fermented liquors ; and a result of it is seen in hard drinkers, in the usually enlarged and softened condition of the organ, and in more or less fatty degeneration.

The exhaustion of functional power occasioned by over-excitement results usually in a reduced size of the hepatic substance (Atrophy), and frequently produces the condition called *cirrhosis*, or hob-nail liver. It would seem that the earlier effects of alcoholic poisoning are to induce inflammation with *hypertrophy*, while atrophy and cirrhosis are a more remote and less frequent result.

Kidneys. — These organs, excited to undue activity by the presence of alcohol in the blood, become impaired, not only as to functional power, but in the very constitution of their tissues. Much morbid matter (*materies morbi*), that it is their office to eliminate and

2

expel, is retained in the general system, giving
rise to gout, rheumatism, &c. The retention
of urea in the blood, owing to the antagonism
between this substance and nervous matter,
gives rise to delirium, convulsions, and tetanus. The kidneys themselves become subject
to various forms of disease, — inflammation,
acute or chronic, hypertrophy, atrophy, &c.
The supervention of Bright's Disease — gran-
ular degeneration — as a new complication,
might, we think, be traced to the abuse of
spirituous liquors; for the organs, at first overworked and weakened, become greatly liable
to the reception of noxious influences; and
what more noxious than those of alcoholic
poisoning can there be?

The effects of intoxicants are also plainly
seen in the diminished power of the inebriate
to sustain inflammatory attacks and local inju-
ries. In such subjects, surgical operations are
attended with great irritation and danger; a
deficient plasticity of the blood greatly retards
the healing process; and any internal organ
affected with inflammation is liable to extensive

suppuration, or to pass into a gangrenous state, with fatal termination; a very slight abrasion of the skin will frequently develop Erysipelas or severe inflammation.

Many cutaneous disorders, especially of the face, resulting from imperfect nutrition, may arise as a consequence of this poison in the blood; and there is no doubt that the skin of intemperate persons is especially liable to carbuncles, boils, &c., as well as to milder forms of eruption. Dr. Darwin mentions a disease called *Psora Ebriosum*, which seems to result directly from intemperate habits.

It is generally conceded, that an intemperate use of alcohol predisposes the system to the attacks of any epidemic or pestilence which may prevail, and fosters the disease when contracted by other causes. Yellow fever, cholera, &c., most frequently find their victims among the habitually intemperate. In such cases, the prognosis is more than usually unfavorable, and the result more frequently fatal.

Numerous other diseases might be mentioned which hold relations, near or remote, to the use of alcohol. This poison may in truth be said to predispose the system to every form of acute disease, especially to those of an inflammatory nature, and to those resulting from disordered or impaired nutrition. It may also be said, that, in many cases where the predisposition to disease from other causes is slight and easily counteracted, the use of this poison will, of itself, be sufficient to excite and develop it.

The effects of this destroying agent are painfully seen in the weakness and general debility of offspring, and are worthy of notice here. When we consider the fact of hereditary tendency to disease, physical and mental, it is reasonable to anticipate that the children of the intemperate will be burdened with many of the diseases which, developed or undeveloped, exist in the parent at the time of procreation. It is well established, and generally believed, that mental debility, insanity, and idiocy, as well as physical maladies, — gout, scrofula, and consumption, — may appear

in the offspring as the result of intemperate
habits, the constant presence of alcohol in
the blood of the parent: more especially is
this true when the habits of both parents are
alike intemperate.

Treatment. — It is very rarely the case that
a single instance of alcoholic poisoning will
lead to any result requiring medical treatment.
It is usually safe to allow the patient to re-
main in the condition of partial coma resulting from intoxication, until the elimination of
the alcohol by natural process of respiration,
secretions of the skin, kidneys, &c., restores
the patient to consciousness and a normal
condition. Medical assistance is usually only
sought after paroxysms of drinking, where alcohol has been present in the blood in large
quantities for several days or weeks, or in
cases where the quantity taken has been less
in amount, but where the use of it has been
long continued and habitual. Anorexia, insomnia, and great nervous prostration, will
follow; and a careful diagnosis will frequently
detect a greater or less development of one or

more of the local or general diseases already mentioned, as liable to be induced by the poison. It is necessary to give a proper tone to the stomach, to produce sleep, and to effect a normal condition of the nervous system as soon as possible. The remedy applied must be such as will produce these effects with as little stimulating or narcotizing influence as possible, and such as will not be contra-indicated by symptoms which may be observed of any accompanying disease.

The first action must be the complete withdrawal of the stimulant, and a substitution of simple and nutritious food that will be attractive and easily digested: amongst these are gruel, beef tea, light soups, &c. The list of drugs necessary to be used in such cases is very small. I have found, in a somewhat extensive experience with these cases, that the remedy best embodying these qualities required, is the *Bromide of Potassium*, given in doses of from twenty to forty grains every four or five hours. In cases of great nervous exaltation, the dose may be increased, and given oftener. This should be well diluted, to prevent irritation of the fauces.

The action of this salt seems to be directly upon the nervous centres, and, acts as a powerful sedative, producing a quiet condition of the system. Its advantage over some other remedies usually given in such cases is, that it does not seem to deplete or weaken the patient; and, as far as my observation goes, it is not contra-indicated, however complicated the case may be. In this it has the advantage over opiates and other narcotics. I have also found it to have great eliminative powers in the removal of *materies morbi* collected and diffused in the system. In reviewing my experience with its use in more than three hundred cases, I have seen no evil effects result from its use, although many of these cases have been very complicated.

As a prophylactic in cases of impending mania, its results have been most satisfactory; controlling, by its anesthetic properties, reflex nervous action, and overcoming, slowly but surely, the exalted susceptibility of the nervous centres. This palliative influence is evidently owing to its positive anesthetic, indirect narcotic power, and direct sedative action.

When mania à potu has actually supervened, its results have been less marked and satisfactory; but yet *I should give it the preference in any form of mania over any drug now in use.*

When the physician has been so fortunate as to see his patient restored to a normal condition, he should remember that his duties are not yet complete. It may safely be laid down as a rule, that no man has thus needed his care that does not stand in need of still further assistance and advice. When the immediate effects of the poison have yielded to treatment, its remote effects, as well as the causes which have led the patient into indulgence, should be carefully diagnosticated, and treated with discrimination and prudence; and while the Materia Medica furnishes drugs, valuable and perhaps even specific for immediate use, it should be borne in mind that no specific has been provided as a preventative of further indulgence. And the complex nature of the disease, varying both in cause and effect with every form of constitution and habit, renders it extremely improbable that any such specific will ever be discovered.

In submitting the following advice for the moral and psychological, as well as medical and hygienic, treatment of methomania, I am aware that it may seem somewhat vague and uncertain to many who would gladly be recommended to some more positive and specific action; but my experience has convinced me, that each case must be approached with rare discernment of individual characteristics and circumstances, and that no one rule of treatment throughout is capable of universal application. I have seen many patients, as I believe, cured of this disease who had been considered hopeless victims to it, and who had so considered themselves; but I have seen no case of complete or even of partial recovery, that could be attributed to the specific action of any drug, or to any treatment that was not substantially the same as that recommended

2*

below, combined with such other treatment as the special requirements of the case seemed to demand. The subjoined advice is the simple result of some years of experience; and, while it is submitted with modesty, it is with the assurance that medical science has as yet discovered nothing more effective. As the effects of this agent are not alone on the physical system, but on the intellectual and moral nature as well, it is necessary that the treatment used should have both a moral and medical character. These patients may be divided into two general classes.

1st, Those who drink from habit or custom, or social influence, and who, from low moral nature or deficiency of self-control, continue the practice until the use becomes inordinate and excessive, and the overstimulated and disabled functions demand the constant excitement of the poison to promote their action; and, 2d, Those with whom this self-poisoning is an intermittent mania, aroused at longer or shorter intervals, and who drink only for the days or weeks that the mania is upon them,

each paroxysm succeeded by a long interval,
perhaps months, of sobriety and abstinence,
but too likely to be followed again by a return
of temptation and a renewed yielding to irre-
,sistible desire. The first we call constant
drinking; the second, periodical.

The patient should first be made to believe,
if possible, that his recovery will probably follow proper treatment. The idea has been so
largely held and advanced by medical men,
that recovery from this disease was impossible,
— in short, that the drunkard could not be
reformed, — that a large amount of skepticism
upon this point has pervaded the community.
And it is a noticeable fact, that those most
likely to adopt this view, and to cling to it with
tenacity, are the unfortunate victims of the
disease themselves. Now, as all psychological
treatment is useless without the earnest and
hopeful assistance of the patient, and as no
sympathetic co-operation can be expected from
one who has no faith in the ultimate success
of his efforts, it follows that the element of
hope should be carefully nourished as a power-

ful stimulant to the other means employed. Medical advisers, as well as the friends of the patient, should remember, that, however discouraging their own experience in similar cases may have been, there are many compe-. tent authorities who claim that recovery is the most likely result of proper treatment, and should endeavor that the patient has none but those views which the most hopeful have entertained.

The patient must then be assured of the physiological fact, demonstrated, I believe, by the discussion of the physical effects of the poison in previous pages, that total and inflexible abstinence from alcohol in any form must be his future rule of life.

In the first class mentioned, a re-introduction of the poison will tend to arrest the healing and recuperative work of nature, and thus rapidly re-induce the precise condition from which the patient has just arisen, while, with the second class, a single indulgence will arouse the sleeping mania to increased and insatiable demand.

It has been my experience, that, except in a few cases of diseased intellectual or moral nature, this important truth, thus sanctioned by medical and scientific authority, and co-inciding as it does with the experience and common sense of the patient, will be generally admitted ; and it thus becomes the foundation on which future action should be based : any starting-point short of this will generally prove a failure. Moral and medical treatment should then combine to confirm and encourage the patient in the continued use of this infallible prophylactic means. It will thus be seen that the result of moral means is not to convince a doubter or a skeptic, but to strengthen a nature morally and physically weakened by this poison, in the presence of a great temptation.

The intimate relations between a normal and abnormal physical state, and a similar moral condition, imply at once, that the first duty of the physician is to assist and direct, so far as his skill will permit, the energies of nature, to the repairing and regulating of such disturbances of the organic or general functions as may have resulted from this poisonous agent.

A normal physical condition, so far as possible, should be sought; and it would be well if the physician would allow his search for symptoms and diagnostic points to anticipate and go beyond the complaints of the patient, that any latent or unsuspected disease may be discovered and treated. The diet should be limited to such food as is nourishing and strengthening, avoiding strong condiments or any thing of a stimulating nature beyond natural food. The patient should be convinced of the fact that he has a special object to gain by the most temperate and judicious manner of life, and that any impeded functional activity is to be especially avoided and guarded against. Of the strictly moral treatment it may be said, that no specific means can be prescribed as having a general application: it should be such as the good judgment of the intelligent physician suggests, as best fitted to the moral and intellectual condition of his patient. It may, however, be generally stated, that it is well to impress upon the patient, by all the means at command, the vital interests to himself that hang upon his total and continued abstinence; to encourage and increase so far as possible his self-respect; and to stimulate into active exercise that family affection and domestic disposition which will of themselves act as a powerful restraint. The usually exciting and demoralizing character of the past life should be avoided; while a habit of mind should be sought, calm, even in temperament, cheerful in disposition, and free from unusual or unnecessary excitement. The patient should be encouraged, that his disease can be cured, and at the same time impressed with the belief that it rests mostly with himself, and that " eternal vigilance is the price of liberty."

It will not, I hope, be regarded in bad taste for me to say, that I regard an institution embodying these views, where the patient may place himself under a voluntary but by no means irksome restraint, as an important adjunct to this mode of treatment.

Prognosis. — The prognosis, I think, is by no means as unfavorable as has been generally regarded. With abstinence from the poison, and such combined treatment as is prescribed above, the adjunctive and resulting diseases, such as renal and hepatic diseases, disorders of nutrition, epilepsy, &c., may be cured ; and, by the removal of these abnormal conditions, the probability of relapse is diminished.

A proper hygienic and moral treatment will then give healthful tone to mind and body, thus re-instating that self-control which is an effectual prevention of the disease. It is in my power to testify to many cases which, I believe, have met with a permanent cure, and to many more where it has been so mitigated that its ultimate cure is regarded as almost certain. In cases where the disease takes the form of mania, and is evidently the result of hereditary predisposition, either from the insanity or intemperance of the parents, the prognosis is unfavorable, and an alleviation of the disease is the only result to be expected. I wish to observe incidentally, that it should be the especial care of the physician, when the effect of the poison has existed in the history of his patient's case, however long past it may have been, to see that no preparation involving an alcoholic admixture is administered or sanctioned by him, as a re-introduction of this poison under any circumstances will almost certainly produce a condition much worse than that attempted to be relieved. Modern science is not restricted to the use of alcohol as an agent for extracting the essential qualities of drugs; and many leading and competent physicians, in view of the possible danger resulting from alcoholic mixtures, are daily prescribing such preparations of drugs as are free from this dangerous agent, as, for instance, fluid extracts instead of tinctures, &c. It should be borne in mind, that the proportion of alcohol is fully seven-eighths of most tinctures. And I have known tinctures of several drugs to be taken in large quantities by patients who had an insatiable craving for alcohol, and to whom the more legitimate means of intoxication were denied.

I think physicians have rarely appreciated the importance of this fact; and a well-known and respected medical author, writing, not long since, upon the very disease of which I am now treating, while recommending the strictest total abstinence, also prescribes tincture of rhubarb, in doses the size and frequency of which would, in my opinion, serve to keep the appetite constantly alive.

I cannot refrain from a word here of warning in relation to the many nostrums or patent medicines with which the community is deluged, the seductive and extravagant advertisements of which stare at one from every available spot. It is a well-known fact, that a large proportion of these preparations, thus widely advertised by unscrupulous proprietors as a panacea for every ill, are nothing but a liquor preparation of the vilest and most adulterated quality, slightly disguised by some common root or herb, and depending for popularity on the most unmitigated puffery and lying, and the fact that they arouse and satisfy the diseased appetites of so many in the community. The frequency with which the weakened and long-controlled appetite is strengthened again to a point beyond control by their thoughtless and injudicious use, is much greater than is generally supposed. And it should not be forgotten, that the disease of which I am writing may be first engendered by their disastrous agency. It is not, of course, expected that the regular physician will countenance them in his practice; but he should make a special effort to discourage their use among those by whom he is consulted, for this reason, if for no other.

It is apparent, I think, that the disease which I have ventured to call *Methomania*, with its complex and varied character, and involving as it does abnormal conditions of both mind and body, must demand of the faithful physician all his resources of physiological and psychological science. He must not be content with merely raising the patient from prostration, but his aim should always be to accomplish, as far as possible, the complete eradication of the disease; and let him, in

every such case, bear with him the consciousness, that, if he succeed in this attempt, the
result of his achievement is to save his patient
from moral and social degradation and ruin,
as well as from physical death.

I have thus endeavored to treat of this
disease in its psychical and physiological developments ; and I cannot refrain, in conclusion, from a few observations on the alarming
prevalency of the disease, the causes which
lead to it, and the duties of the community
respecting it.   It is a sad fact, that, while to be
afflicted thus personally is one of the greatest
misfortunes that can befall a man, the suffering occasioned by it cannot be endured alone,
but must be shared with those whose happiness
is often dearer to the victim than his own ;
and who, blameless themselves, are powerless
to free the sufferer, or lift the burden from
their own hearts, but can only pray with such
faith as they have, and hope with weary long-
ing for that alleviation which so often never
comes.

My own intimate relations with this class of

patients for many years, has opened to my view thousands of hearts locked, with proud sadness, from the intrusive sympathy of the world at large. And I believe to this cause alone may be attributed more of mental anguish, of crushing anxiety, of the sickness of hope deferred and often blasted, of death even, that makes of the overcharged heart its easy prey, than to any other of the causes responsible for human misery.

It seems to me that the very general prevalency of the disease is not appreciated. Every one knows of one or more cases, but they are looked upon as infrequent exceptions; whereas it may be said that there is hardly a family of any size which has not amongst its members at least one who is either a confirmed inebriate, or more or less tainted with this disease. It must be borne in mind, that a large proportion of these cases are not visible to the eye of society, but are hidden from the world by the decent pride of friends, or the sensitive modesty of the patients themselves. No social limits are narrow enough to exclude it, and it knows no distinction of sex. It is the skeleton that sits at every board, and darkens with the gloom of its presence the .brightest scenes of life.

Does it not, then, become the duty of all to ask themselves, How far am I, or is society through me, responsible for the alarming prevalence of this disease? and can any conduct or influence of mine mitigate the evil? I am not simply discussing the temperance question, and will not repeat all the old arguments used by the leaders of that reform.

The public are already familiar with them, and fully appreciate the necessity, from prudential as well as moral reasons, of arresting the gigantic evil of intemperance. The only issue fairly raised in the community is on the efficacy of the various means proposed. I do not wish, in this simple treatise, by endorsing any scheme, to take issue with the advocates of any other, but simply to utter a caution against courses most likely to develop this disease, or to retard or prevent the recovery of those already affected with it. I therefore consider it the duty of every one to discourage as much as possible the use of alcohol in the community. I am aware of the strong hold it has on society as an agreeable stimulant and a pleasant element of social life.

These features have enlisted on its side the subtle and almost omnipotent power of fashion, which, in our land, as almost everywhere else, is one of the strongest of social forces, and the hardest to resist. But when we consider the fearful nature of the disease, and the fearful consequences it entails on the health and fortunes and happiness of the victim as well as on those who love him, and who perhaps look to him for support and protection, its evil influences not content with one generation, but spreading its malign taint among our remotest descendants, and darkening the future of our children's children; and when we consider that only by indulgence in the poison can the disease be engendered or developed, — should not every intelligent and conscientious man ask himself, Shall I yield my better judgment in weak subserviency to unreasoning fashion? or shall I take counsel of my nobler nature, and discourage by act and influence the use of alcoholic stimulants in society? I think there can be no doubt on which side the weight of argument lies; but, if any man doubt or hesitate to make the sacrifice, I would remind him of the well-known and prudent commercial maxim, that it is worth something to insure a risk, and let him remember, that while indulgence at the best is uncertain, and carries with it a fearful risk, total abstinence is a protection as sure and undoubted as the oracles of God. I have already alluded to the evils often resulting from alcoholic mixture administered as a medicine; but I would again warn the friends of any liable to this disease, or who have been restored from it, that they guard with especial care this much-exposed point of attack. Let no false pride or delicacy prevent them from confiding, in any physician they may consult, so much of the history of the patient's case as will show him the danger of an alcoholic prescription, and I feel sure that in such a

case any good physician would take counsel
of his conscience and his prudence, and avoid,
if possible, that agent in the preparation of
his remedies. I wish to offer one other suggestion, especially to those who are brought in
contact in any way with the sufferers from
this disease ; and that is, that, in dealing with
such a case, they act from a proper appreciation of the sufferings of the patient and his
struggles to regain a normal state. I have no
desire to apologize for gross intemperance, or
to attribute it to an uncontrollable force or
involuntary action on the part of the inebriate. None are less willing to so consider it
than the victims themselves, who know that
self-control, though weakened, is not wholly
lost. But, while I would not condone a vice,
I would give a sympathetic assistance to
weakness, and a kind encouragement to good
resolutions.

Let it be remembered, that such a man is
diseased, and that he is fighting not against
temptation only, but against temptation fos-
tered and encouraged by the morbid elements

3

of his own physical and mental nature. In such an unequal contest, let our sympathies go with the higher and holier aspirations, against the lower and diseased nature; and, while we apply as far as possible the treatment enjoined in previous pages, let us see to it that no act of ours shall counteract their influence, or throw darkness or discouragement in a pathway so difficult and obscure. How often may it be said of those who are struggling with almost hopeless effort to rid themselves of the "body of this death," that "their foes are they of their own household"! Such conduct on the part of friends is of course mistaken, — often from mistaken kindness; but it has none the less been a fatal blow to many an honest effort to reform. I would therefore counsel, in such cases, the utmost patience and forbearance on the part of friends; not that weak pity that discourages effort, but a generous tolerance mingled with firmness; such encouragement as shall stimulate exertion, and a confiding love that shall draw the patient by the higher motives of his being. I cannot refrain here from suggesting the exercise of some religious influence and example. Of course I cannot be expected to indicate its nature. It must be decided by those who employ it. But it has been my custom to present each patient of my own, after recovery, and when about entering society with its unavoidable temptations, with a copy of the Bible, and the advice to read daily of its sacred pages. Aside from the moral elevation that must follow the reading of God's word, I think the act of reading serves another end, by daily reminding the reader of his pledge to abstinence, and thus daily invigorating his will for the contest with temptation. I can remember no single instance of a relapse where my advice in this respect has been followed; but I can recall scores, once sunk low in intemperance, who will attribute their salvation mainly to this means.

Public sentiment has been much modified of late on the question of the proper treatment of the inebriate; but it is certain that

the subject is as yet but imperfectly compre-
hended. It is not presumed that the subject
is exhausted in these pages : they merely em-
brace a few of my own observations from a
somewhat lengthy and extensive intercourse
with patients of this class. No thoughtful
man can be satisfied with the present achieve-
ments in the treatment of this disease. Let
us hope, that, by careful study and observation,
many truths may yet be developed that shall
mitigate the disease, and add to the chances
of recovery. At present, we may truly be
thankful that a reform in this matter has
commenced.

It is not many years since no thought of
humanity entered into the treatment of the
insane. Manacles, dungeons, and scourges
were the only instrumentalities thought fit to
be enlisted by the wisdom of two generations
ago ; but a later and more humane civilization
has so ameliorated their condition, that the
utmost kindness, consistent with their own
and the public safety, is now demanded of
those having them in charge. I look for a

similar revulsion of feeling in the treatment of the inebriate, as a result of sympathetic appreciation and intelligent judgment; and, under it, we may expect to achieve much greater success in our efforts in their behalf.

APPENDIX.

BY PROF. H. R. STORER, M.D., OF BOSTON.

My friend, Dr. Day, has requested me to add a word to those he has said in the previous pages. This courtesy of his I presume to be owing to the pleasant personal relations I held to him, as an assistant instructor in the Medical School of Harvard University, during a portion of his pupilage, and the many conversations we then had together upon the medical, medico-legal, and social aspects of the dread disease to whose cure he is now devoting his life. Be this as it may, I take pleasure in acknowledging alike the excellence of the treatise he has written, its strictly philosophical spirit, the practical influence it must have upon the community, and, above all, the truly scientific manner in which ebriety and its ef-

[55]

fects, almost for the first time in the history of medicine, are now being treated at his hands. The flattering encomiums he received from Miss Catherine Beecher, almost by chance as it were and to fill out the measure of a humorous story in a late number of our most extensively circulated periodical,* were not at all undeserved.

It is alike the glory of Massachusetts and of our own profession, that almost the first steps, upon an extended scale and by public authority, towards the proper estimate of drunkenness as a disease, and its cure by appropriate medication and physical as well as moral hygiene, have been taken in this city. The State of New York, it is true, has spent enormous sums upon the Asylum at Binghampton; but this, from alleged mismanagement, is said thus far to have proved scarcely more than a splendid failure.

Some three years since, my attention was

---

* "The Little Black Dogs of Berkshire," *Harper's Magazine, August,* 1866.

more forcibly directed to this subject, while acting as one of the State Commissioners on Insanity, then appointed by the Governor and Council, to examine into the condition of the lunatic asylums of this Commonwealth, and the system under which they are administered. Among the many topics that presented themselves to the attention of the Commission were, very naturally, the causation and cure of the several varieties of mental derangement, and whether these questions had as yet been definitely settled, or exhausted either of scientific or practical interest. It is needless to say, that my colleagues — Josiah Quincy, Esq., of Boston, and Alfred Hitchcock, M.D., of Fitchburg — were, with myself, satisfied that the work has been but just begun : and that while our American asylums, more especially those of New England, — for we instituted a very extended personal comparison of our own with those of distant States and the Canadas, — are in many respects all that can be desired ; in other respects, — and the remark covers certain points, or I might rather say the general system, of medical management, — they are still far behind the hospitals of all other branches of applied medicine. The cells and chains and stripes of former days are gone; but by a very natural transition, in simply the most perfect kindness of treatment, the most quiet seclusion, and the most complete protection from ordinary exciting causes, it is sought to soothe an unruly mind, or to interest an apathetic one, both being simply diseased, and diseased, in a large proportion of cases, through reflex action, by sympathy with some organ distant from the brain; this being especially the case with women, as I have elsewhere shown,* the brain being always the seat of the insanity, but not always the seat of its cause. The Commission found, that while a large proportion of the crimes that are committed are the effect, direct or indirect, of intoxication, and that while inebriety under certain circumstances, and under the several forms of mani-

---

* The Causation, Course, and Treatment of Insanity in Women. Transactions of the American Medical Association, vol. xvi. 1865, p. 120–255.

acal result described by Dr. Day, clears or should clear from criminal responsibility; yet, at all our asylums, inebriates, of whatever position in life, were either refused admission, or else received with great reluctance. Something was clearly necessary to be done, alike for the sake of the unfortunates themselves, and to protect society. The members of the Commission were therefore unanimous in recommending to the Legislature a still greater confidence in Dr. Day's most excellent institution than had been hitherto evinced, and a more liberal appropriation for its support.* I have been glad to realize, that the reports referred to have been instrumental in accomplishing the desirable result which has since followed.

Reference has been made by the doctor to the dire effects, so often seen by medical men, in the persons of the children of those addicted to habits of intoxication,— epilepsy, idiocy, and

---

* Senate Document No. 72, Feb. 1864, and Senate Doc. No. 1, 1864; the latter being the especial Report of Dr. Hitchcock after visiting the Asylum at Binghampton.

insanity, congenital or subsequently developing themselves, with or without any apparent exciting cause. He has not, however, I think, sufficiently held up to the victims of this baleful thirst the terrible curse they thus deliberately entail, or may entail, upon their descendants. It is not merely the man or woman inflamed by alcohol, in any of its crude or dainty preparations, at or near the time of sexual intercourse, that implants the fatal thorn in the child at the very moment of its conception. They are equally guilty, perhaps even more so, who — their blood diseased from long saturation with this poison, their nervous system shattered, the very fountains of their being tainted — proceed, whether deliberately or inconsiderately, to engender offspring. — "Woe unto the children of drunkenness," &c.

The caution has been impressed upon physicians, never in any way, by their medicinal preparations, to re-awaken a thirst that, once raging, has been assuaged and put at rest. Dr. Day's remarks upon this point are admirable. I remember with grief a case occurring some thirteen years ago, shortly after my entrance into practice, where I was thus the unintending author of a great deal of misery. There is another phase of the danger, however, equally important, which has not as yet been alluded to. I refer to the original awakening of the passion for drink at the hands of medical men. If it is a misfortune to cause those to fall who have already once lost their self-respect, far worse is it, in some respects at least, to be the means of bringing an hitherto healthful mind to grief. Yet this is very constantly done, when, by the physician's order, tonics are administered to the weak, and alteratives to the diseased, the basis of which is alcohol. To the maxim, that drunkards do not generally die of consumption, is undoubtedly due the fall, and the subsequent 'drunkard's death, whether by a cirrhosed liver, in an affray, or by the gallows, of many a person threatened with, perhaps even attacked by, pulmonary disease. Whiskey, fusel-oil, baths by alcoholic vaporization or amid the fumes of a distillery, alike serve, to the timid, as a shield against all but the stab from within; to the weak, as a staff more treacherous than a broken reed; and to the already thirst-smitten, as the longed-for excuse to more freely feed the consuming flame. I do not mean by this that the measures referred to ought never to be employed; but they should be resorted to with caution, and with due regard to the needs of each individual patient.

In the case of women, there is a still more horrible risk. A debauched woman is always, everywhere, a more terrible object to behold than a brutish man. We look to see them a little nearer to the angels than ourselves, and so their fall seems greater. In the exercise of my profession, during the first seven years of which I have been especially, and for the subsequent six years exclusively, devoted to the diseases of women, many sad instances of habitual drunkenness, and the affections arising therefrom, have presented themselves to me. In a very large proportion of these cases, just as obtains in the inordinate use of narcotics and anæsthetics, the morbid habit arose

from the liquor having first been prescribed
by some physician for the relief of pain, or by
some relative or nurse to whom similar advice had previously been given. There is no
doubt that the nervous system of woman is
much more exquisitely sensitive to suffering
than that of man ; there is no doubt that they
have infinitely oftener causes for suffering
than do we. It is just as true that they often
bear uncomplainingly, and for years, anguish
of which we have scarcely a conception, and
from a tithe of which the strongest of us
might well shrink. In the throes of dysmen-
orrhœa and of parturition, the lancinating
pains of cancer, the lassitude of lactation and
of early pregnancy, the faintings of hysteria,
and the fatigues following marriage, many
physicians have heedlessly seen fit to order
alcoholic stimulants, not for the moment alone,
but for considerable periods of time. Such
is still done by some obstinate or over-con-
servative men, to the ruin of their victims, to
their own eternal disgrace, and, may we hope,
to their eventual remorse. Cases of the kind

referred to have been frequently sent to me for treatment, or I have been called in consultation to them. Here the exciting cause of their sad condition was an attempt to relieve weakness or to still a pain. But pain and weakness are not in themselves generally the disease itself: they are but its symptoms; and to prescribe for them alone, as is still too frequently done in the case of sick women, without ascertaining in what the actual disease which excites them consists, is alike unscientific, empirical, and inexcusable. To this fact I have more than once called the attention of the profession: in exciting to it the anxiety of women themselves, and of their protectors by blood or by bond, I am but fulfilling the physician's highest duty.

One word more, and I have done. True to the special work to which I have long devoted myself, it is of female inebriates, rather than of men, that I would again speak. A man addicted to intemperance is, until cured of his infirmity, a curse indeed to those about him. A drunken woman would have made purgatory

of Eden: she would make such of heaven.
As yet, we have in New England no retreat
for these unfortunates, suffering themselves
untold-of tortures, worse often than those their
friends are compelled to bear. The necessity
of a special provision for such I some time
since impressed upon my fellow-practitioners,
by a public appeal,* and thus endeavored to do

---

* I append the notice then issued in the "Boston Medical and Surgical Journal," with the editorial comments by
which it was prefaced. The institution itself was given up
from a belief that it should be rather a public than a private work.

"The subject of the following communication is .one the importance of which can hardly be exaggerated. The unfortunate
victims of the vice to which it refers are among the most pitiable
objects to which our professional sympathies are ever directed.
The experience of nearly every physician must have furnished
him with cases of this kind of the most embarrassing character.
Household restraints and home influence are little better than
worthless in these cases: and the prospect of an asylum where
they can be received and tenderly cared for will bring an inde-
scribable relief to many. We have no means of knowing how
extensive a provision is required for the purpose in our own
State; but we hail the commencement of this movement with the
greatest satisfaction, and hope it may meet with the signal success which it deserves.

"RETREAT FOR INTEMPERATE WOMEN. — The necessity of
making some special provision for the victims of intemperance,
partly for the benefit of the individual and partly for that of the
community, is beginning to attract general attention; and the sub-

somewhat to meet the evil. Private action, however, is necessarily but temporary and imperfect; and the work can only be done by the public, whom it most concerns. To rouse

---

ject, in its various bearings, has been brought before the Massachusetts State Board of Commissioners on Insanity, as among the matters deserving their serious consideration.

" Aside from the question of establishing a public asylum for inebriates, the advantages of which would be more naturally confined to the middle and lower classes, it appears that there is as yet in New England no place of refuge for intemperate women of good social position, except the public and private lunatic asylums, which are unfitted, in the almost unanimous opinion of their superintendents, for the reception of such cases; at many asylums, indeed, admittance being refused to them, alike in justice to the other patients and to the inebriates themselves. The number of applications at the New-York General Asylum at Binghampton far exceeds the possible capacity of the building; while the Washingtonian Home in Boston, whose influence for good is already so extended, is for men alone.

" In accordance with this apparent want, arrangements have been made by which there will be afforded to a limited number of self-indulgent women, whether addicted to opiates or stimulants, the necessary elements for their cure; namely, voluntary seclusion from temptation, the strictest privacy if desired, a location in the immediate vicinity of the city and yet unrivalled for purity of atmosphere and beauty of scenery. The house selected for the purpose is one constructed with especial reference to a comfortable residence during the winter; attendants will be provided of unexceptionable character, and but few patients will at present be received. For further information, application may be made to the Secretary of the Commission, Dr. H. R. Storer, at Hotel Pelham, Boston; the other members of the Board being Hon. Josiah Quincy, Jr., of Boston, and Dr. Alfred Hitchcock, of the Governor's Council, of Fitchburg. It may be stated, that the step now

the world to a sense of certain necessities, one must handle a very long lever. There is none so long and so effectual as our love for woman, save, indeed, her love for us. To assist in the firmer establishment of the Washingtonian Home, and the extension of its benefits, is a privilege which we well may value. The work will never be completed, however, till wards are provided, or a branch institution built, for inebriate women. Then, the task well rounded, its initiator and the real master workman, Dr. Day, will indeed be blessed.

HOTEL PELHAM, 15th January, 1867.

---

taken has the cordial approval and endorsement of His Excellency Governor Andrew, Judge Hoar of the Supreme Court, Drs. James Jackson, Jacob Bigelow, John Jeffries, H. I. Bowditch, J. Mason Warren, Tyler of the Asylum at Somerville, Jarvis of Dorchester, and other of our more prominent citizens."

GLOSSARY OF SCIENTIFIC TERMS.

ATROPHY, a wasting away.

ANÆSTHETIC, deadened or impaired feeling.

ASPHYXIA, without pulsation.

ANOREXIA, want of appetite.

BRIGHT'S DISEASE, several forms of kidney disease.

CIRRHOSIS, a yellow, contracted, fissured condition of the liver.

CUTANEOUS, pertaining to the skin.

COMATOSE, insensible.

CAROTIDS, large arteries of the neck.

CEREBRITIS, inflammation of the brain.

DEGENERATION, a diseased alteration.

DIAGNOSIS, distinguishing one disease from another.

EPILEPSY, a disease with convulsions and unconsciousness.

ENCEPHALON, the contents of the cranium.

FAUCES, throat and back of the mouth.

FEBRILE, pertaining to fever.

GASTRITIS, inflammation of the stomach.

HEPATIC, pertaining to the liver.

HYPERTROPHY, enlargement of a part from increased nutrition.

INSOMNIA, sleeplessness.

MENINGITIS, inflammation of the membrane of the brain.

PLETHORA, redundancy of blood in the system.

PARAPLEGIA, palsy of the lower half of the body.

PATHOLOGY, the nature and cause of disease.

PROPHYLACTIC, preventing.

PROGNOSIS, judgment of the course or termination of a disease.

RENAL, pertaining to the kidneys.

SUPPURATION, the process of forming pus.

STRABISMUS, want of parallelism in the position and motion of the eyes.

TRACHEA, the wind-pipe, air-passage of the lungs.

TETANUS, a disease characterized by continuous spasms of muscles.

UREA, the nitrogenous constituent of urine.

VASCULAR, pertaining to blood-vessels.